CELLS

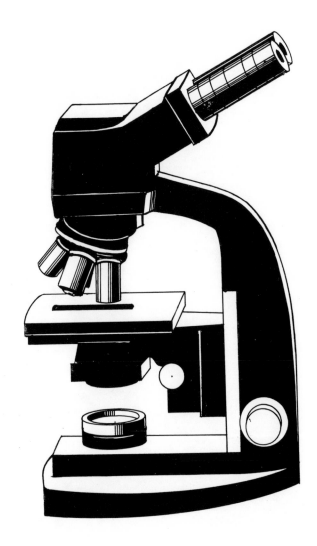

CELLS

The Basic Structure of Life

By Vicki Cobb

illustrations by Leonard Dank

←A FIRST BOOK→

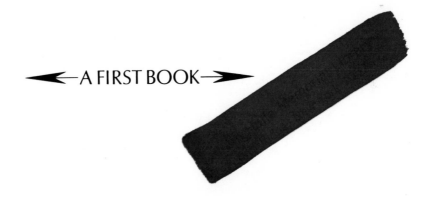

FRANKLIN WATTS | NEW YORK | LONDON

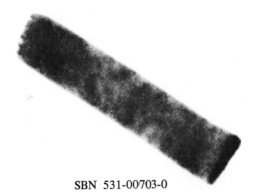

SBN 531-00703-0

Copyright © 1970 by Franklin Watts, Inc.
Library of Congress Catalog Card Number: 79-101745
Printed in the United States of America
5 6

CONTENTS

CELLS

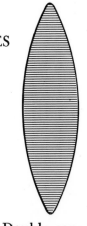

Double convex

Discovery of Cells

Curiosity is a characteristic that is a part of every human being. All of us want to know about ourselves, about life, and about events that affect our lives. We want to know what is yet unknown, to see what cannot yet be seen, and to understand what is still mysterious. And man's curiosity has taken us far toward finding some of the answers. Think of how different life is today from the lives of people in the past. This difference is largely due to curiosity about the unknown, leading to investigation that reveals new information. This new information leads to changes in everyday life, and gives us tools to probe further into the unknown.

Converging lenses used as simple magnifying glass

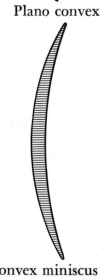

Plano convex

Convex miniscus

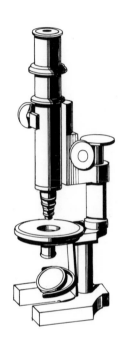

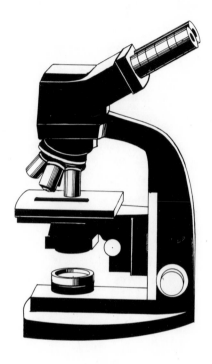

ROBERT HOOKE'S
COMPOUND
MICROSCOPE

(about 1665)

COMMON
MICROSCOPE

(about 1915)

TODAY'S
MICROSCOPE

The discovery of magnifying lenses in the seventeenth century opened up a new world for curious men to investigate. The new world was made up of structures too small to be observed with unaided eyes. Magnifying lenses are transparent objects that are curved outwardly on at least one side. They can be used singly, as a magnifying glass, or in combination with other lenses to produce an enlarged image of a natural object. Early magnifying lenses, with a place to mount a specimen, were called *microscopes*. And early microscopes looked very different from the ones we use today.

There were many technical problems with the first microscopes. For one thing, the lenses were not perfect. In order to magnify natural objects even as little as five times their normal size, early investigators had to look at specimens through dark rings and distortions which were due to the imperfections in their lenses. Some people tried to correct the defects in the lenses by using the imperfect lenses in combination with others. While lenses in combination can give a higher magnification than a single lens, unfortunately the already existing errors are also multiplied.

Considering the problems connected with microscopes used in the seventeenth century, it is amazing that so much was learned from these crude instruments. In 1665, an Englishman named Robert Hooke (1635–1703) published a paper containing his observations with a microscope. In it he described a slice of cork, which is the layer of wood just inside the bark of a tree. Cork seemed to be made up of "walls" surrounding empty spaces. Hooke called these spaces cells, because they reminded him of the small rooms, also called cells, found in monasteries.

Although Robert Hooke was the first to use the word "cells" to describe the structure of cork, the real seventeenth-century giant

3

in the field of microscopes was a linen merchant in Delft, Holland. His name was Anton van Leeuwenhoek (*lay*-ven-huke), and he owed his success to his lenses, which he ground himself. In fact, he had the best lenses in the world. Because he was such a perfectionist when grinding lenses, his lenses did not contain as many errors as those made by others.

Leeuwenhoek (1632–1723) looked at anything and everything he could think of. A list of his specimens reads like a recipe for a witch's brew. He looked at hair and blood and stagnant pond water. He described and drew pictures of "animicules" and bacteria. For over fifty years, Leeuwenhoek shared his observations with other

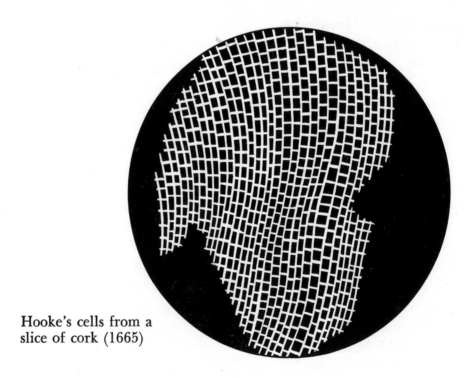

Hooke's cells from a
slice of cork (1665)

LEEUWENHOEK

scientists by sending papers to the Royal Society in London, a scientific organization founded in 1660. Unfortunately, however, he did not share his methods. Therefore, no one else saw much of what he had seen until over one hundred years after his death.

Today we think of cells as the basic structure of all forms of life. But it took a long time for scientists to come to this idea. Part of the problem was in the word "cell" itself. When Hooke used the word in his description of cork, he drew attention to the walls

CELL AS SEEN IN MICROSCOPE

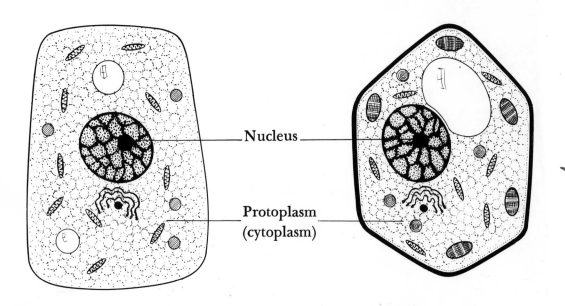

Nucleus

Protoplasm
(cytoplasm)

Animal cell

Plant cell

which formed the cells. Only after many years, and many descriptions of all kinds of living things, did people begin to realize that the importance of cells was not in the walls (which are not even found in animal cells), but in the contents of the cells themselves.

All cells in both the plant and animal kingdoms contain a jelly-like, living substance called *protoplasm* (*pro*-toe-plasm), meaning "the first thing formed." This substance was considered to be alive because it moved and it could reproduce itself. Included within the protoplasm of most, although not all, cells is a small body called a *nucleus* (*new*-klee-us; plural, nuclei — *new*-klee-i), which means "central body." Scientists also use the word *cytoplasm* (*sigh*-toe-plasm) for all cellular protoplasm except the nucleus.

Now, scientists focus their attention on protoplasm and nuclei, hoping to learn more about the secret of life. Although we know a great deal about cells today, we are still trying to answer basic questions raised by nineteenth-century biologists: What is the nature of protoplasm? What is the role of the nucleus in cell activities?

Where Are Cells?

Life comes in many forms. In fact, there are hundreds of thousands of different kinds of plants and animals. Early biologists devoted their time to making some kind of order out of the confusion found in all the varieties of life. Their job was to sort out species of plants and animals that had similar characteristics. In other words, they spent their time classifying plants and animals.

Early microscope workers were also looking for order. They hoped to find something that all living things had in common.

PLANT CELLS

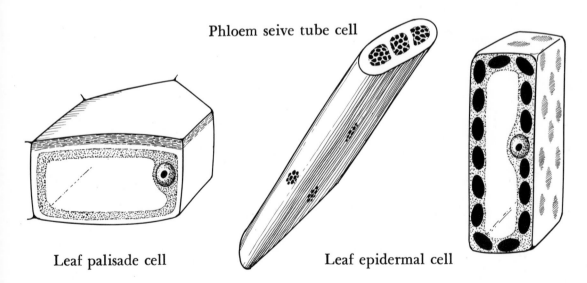

Phloem seive tube cell

Leaf palisade cell

Leaf epidermal cell

8

And they hoped that they could find this common ground on the microscopic level. Although they wanted to look at every type of life under their microscopes, they soon encountered all sorts of problems. For one thing, in order for a specimen to be seen under a microscope, it must be thin enough to allow light to pass through it. Most living things that we see around us are not this thin. So, techniques of slicing, peeling, and squashing living material had to be developed to overcome this problem. Then came the problem of the amount of detail revealed by microscopes. Leeuwenhoek saw the most detail for his time when he achieved a magnification

ANIMAL CELLS

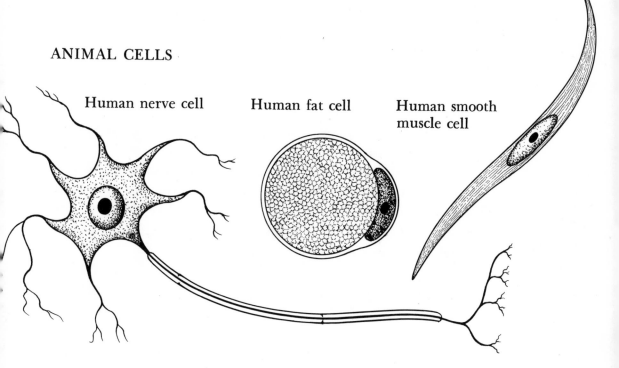

Human nerve cell   Human fat cell   Human smooth
                                        muscle cell

9

of 200 times actual size. Today a student microscope gives, at the least, a magnification of 400 times actual size, and some give up to 2,000 times actual size. You can see more with a modern ordinary microscope than with the best microscope made before 1830.

Nevertheless, as the observations of early investigators accumulated, it became more and more clear that there *was* some order in the microscopic world. Every living thing seemed to be made up of smaller structures that had distinct boundaries. These structures (which, as you know, came to be called cells) usually had nuclei, and were of different shapes and sizes. Plant cells had rigid cell walls, while animal cells were contained in thin membranes. Some held colored material.

Despite these differences the scientists, no matter how hard they looked, could not find a single living thing that was made up of less than one cell. Eventually, these observations led scientists to make a general statement that has become part of modern cell theory: *All living things are made of cells.*

But, in the strange way of science, as soon as one important question is answered, all sorts of other questions crop up. We learned that all living things were alike in the sense that they were all made of cells. Now we had to learn the answers to such questions as: What do different kinds of cells look like? How do cells stay alive? How do cells develop? How does a cell in a plant or animal do its special job?

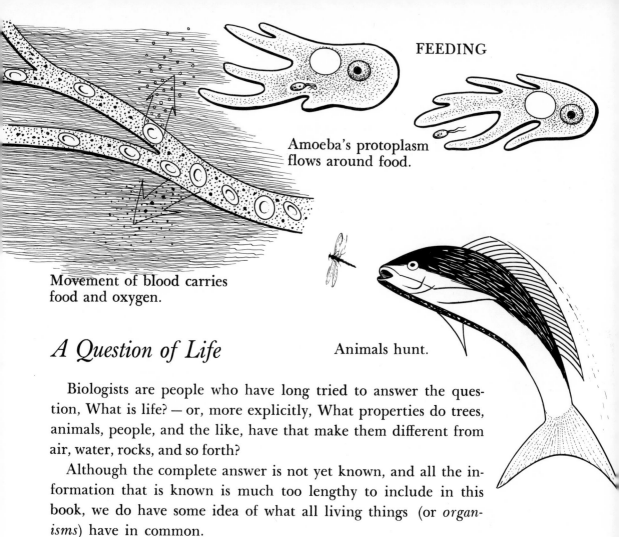

FEEDING

Amoeba's protoplasm
flows around food.

Movement of blood carries
food and oxygen.

A Question of Life

Animals hunt.

Biologists are people who have long tried to answer the question, What is life? — or, more explicitly, What properties do trees, animals, people, and the like, have that make them different from air, water, rocks, and so forth?

Although the complete answer is not yet known, and all the information that is known is much too lengthy to include in this book, we do have some idea of what all living things (or *organisms*) have in common.

FEEDING

All organisms must continually be provided with materials we call food. If the food supply is cut off for an extended period, the organism dies. Feeding in animals involves the taking in and breaking down of other plants and animals. Feeding in green plants takes a different form. Here food is made within the plant from

Flowers move
toward sunlight.

EGGS

SEED

EGGS

simple molecules of carbon dioxide and water. But once the food is made, it is used the same way in plants as in animals to keep the organism alive.

In order to feed, living things must be able to move. Movement can take many forms. It may be the large and obvious movements of running, flying, or swimming animals. It may be the less obvious movement of blood circulating through your body. It may be very slow, like a morning glory turning toward the sun. It may be very small, like the protoplasm streaming through the cells of a leaf. This kind of motion can be seen only through a microscope.

COORDINATION

Almost all living things change with time. This change is very dramatic in many forms of animals and plants. Think of the difference between an elm seedling and an elm tree. Caterpillars show almost no resemblance to the butterflies they become. It is almost impossible to imagine how a human being can develop from a single fertilized cell. These kinds of changes are called growth and development.

Movement, growth, and development are examples of the coordination found in living things. Each part of an organism is somehow "in touch" with the rest of it. Each activity regulates other activities. An example of what happens when some coordination breaks down can be seen in a disease such as cancer, which we will discuss later. Here, growth is wild and uncontrollable, and without communication with the rest of the organism.

REPRODUCTION

Perhaps the most outstanding characteristic that distinguishes the living from the nonliving is that living things reproduce. Germs produce more germs, cats produce kittens, asparagus plants produce more asparagus plants.

Reproduction means that no matter what fate any individual of a species may suffer, the life of the species continues. The death of a dog does not mean that there will be no more dogs. Every person does not have to have children in order for the human race to survive.

Requirements for Life

The activities we have just outlined are not possible without energy and raw materials for construction. Food obviously provides the raw materials. In most forms of life, the source of energy is the combining of food with oxygen. This process is called *oxidation*. Because you have seen fire, which is the rapid release of heat and light energy as fuel combines with oxygen, you know that oxidation produces energy. In living things, the oxidation of food is carefully controlled so that energy is not released all at once, as is the case with fire, but is released in small bits as needed for various activities.

The reason for so much talk about animals and plants in a book about cells is that: *The basic activities and requirements of plants and animals are also the basic activities and requirements of cells.* Cells require food and oxygen. One activity within a cell must be coordinated with its other activities. When cells associate with other cells, the activities of each cell must be coordinated with what other cells do. Cells also reproduce, grow, and change.

Let us look at some of the various ways in which the different kinds of cells do all these important jobs to stay alive.

Cells That Live Alone

Cells that live alone come in a variety of shapes and sizes. Some belong to the plant kingdom, some to the animal kingdom, and some seem to have the traits of both plants and animals. They all, however, have one thing in common: No matter where they live, they do not depend on each other for survival. Like true pioneers, cells that live alone get their own food and provide their own defenses against their enemies. Biologists consider them to be simple organisms because they are single cells. Yet, even these one-celled organisms are so complicated that we do not fully understand how they perform all the activities necessary for life.

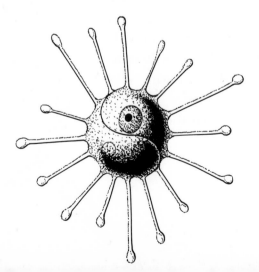

COCCUS

Bacteria

Bacteria are among the smallest living cells, and are considered to be members of the plant kingdom because they have cell walls characteristic of plants. There are three main types of bacteria, classified according to their shapes: coccus — sphere-shaped; bacillus — rod-shaped; and spirillum — spiral-shaped.

In the course of evolution, bacteria have developed that can exist on almost any kind of substance which contains the element carbon. Bacteria live on leather, rubber, and petroleum, in addition to the kinds of plant and animal material that we use as food. Some bacteria use oxygen from the air. Others can exist in an environment free of oxygen gas. Many kinds of bacteria are unable to move from one place to another. Some have hairlike extensions of protoplasm called *flagella,* that act as whips in moving bacteria through liquids.

BACILLUS

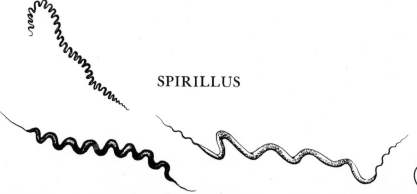

SPIRILLUS

The Amoeba

The amoeba is one of the largest and most primitive of a group of one-celled animals called *protozoa*. It lives in pond water, and it uses a strange kind of movement for both locomotion and food-getting.

An amoeba is an irregular blob of protoplasm that forms extensions of itself called *pseudopods* (*sue*-do-pods), meaning "false feet." The rest of the amoeba flows into the pseudopod, and in this manner the entire amoeba has moved. Amoebas "trap" food by extending pseudopods around a bit of material. The pseudopods meet and join, forming an empty-looking body, called a *food vacuole,* around the material. (*Vacuole* comes from a word meaning "vacuum" or "empty.")

Chemicals formed by the amoeba digest the food while it is in the food vacuole. The substances which the amoeba uses are absorbed through the thin vacuole membrane. The undigestible material is expelled along with the food vacuole. In addition to a nucleus and food vacuoles, an amoeba contains another very obvious structure called a *contractile vacuole*. Its job is to expel fluid wastes by quickly contracting.

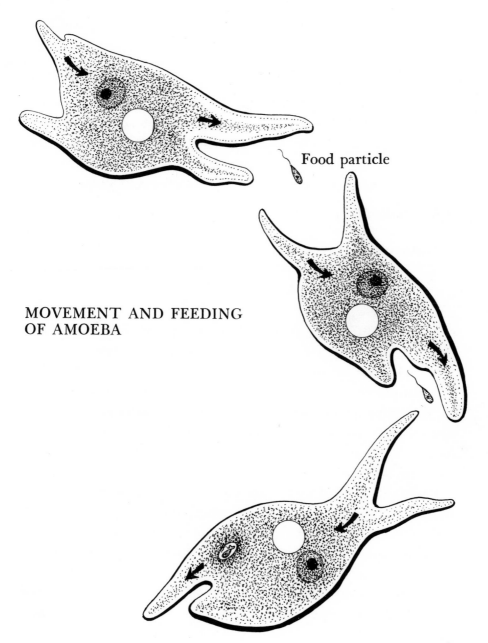

Food particle

MOVEMENT AND FEEDING
OF AMOEBA

Algae and Diatoms

There are many forms of one-celled plants that live in both fresh and salt water. They contain the green pigment *chlorophyll*, found in bodies called *chloroplasts,* and are able to make their own food, like other green plants, from carbon dioxide and water. Primitive forms of algae live alone. Diatoms are one-celled plants with a glasslike skeleton. They come in a variety of shapes.

ALGAE AND DIATOMS

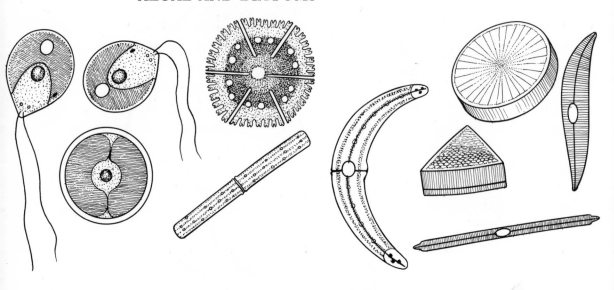

Cells That Live in Groups

What do you think are the advantages and disadvantages of living in a town or city? One advantage is that by living in close proximity with others each individual is given greater protection. A wolf, for example, cannot attack fifty men at the same time. A second advantage is that living with others allows for a division of labor. Some men build houses, some grow food, some treat illnesses, and so on. Such a division of labor lets each man become more skilled and more efficient in his own work. One possible disadvantage to group living is that each individual becomes more and more dependent upon others. If for some reason he is cut off from society, he may not be able to survive.

The advantages and disadvantages of living with others also apply to cells. Cells that associate with other cells are more specialized to do different jobs, and are less independent. There are some cells of animals and plants that show the beginnings of specialization in their association with other cells. They perform special jobs when living with other cells. But they can also live independently until they reproduce to form a new colony. These cells can be compared to a frontiersman who performs a special job when he lives in town but is still pretty good at making his own way in the wilderness.

The Volvox

The volvox is a colony of cells. The number may be between a few hundred and 40,000. The cells are arranged in a single layer, connected to each other with strands of protoplasm, to form a hollow ball filled with a jellylike substance. All the cells have two flagella at the back end and an eyespot at the front. The entire colony moves through water with a rolling motion as the cells whip their flagella. This motion is coordinated through the strands of protoplasm.

The volvox colony shows the beginnings of cell specialization. The cells in the front of the colony have larger eyespots than the cells toward the rear. The eyespots are sensitive to light, and the forward cells determine the direction in which the colony moves. The volvox contains chlorophyll and it makes its own food by using energy from sunlight. So you can see why it is important for a volvox to move toward light. A few cells at the back end of the colony have the special job of reproduction. New volvox colonies are formed inside the parent colony. They are released when the wall of the parent colony splits open.

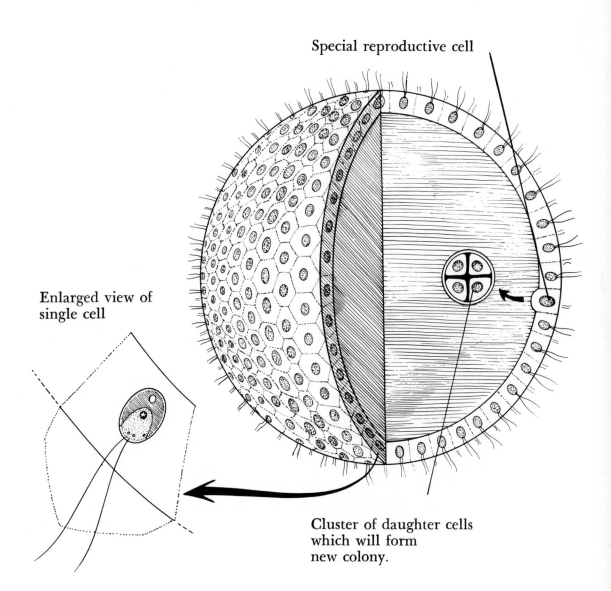

Special reproductive cell

Enlarged view of
single cell

Cluster of daughter cells
which will form
new colony.

23

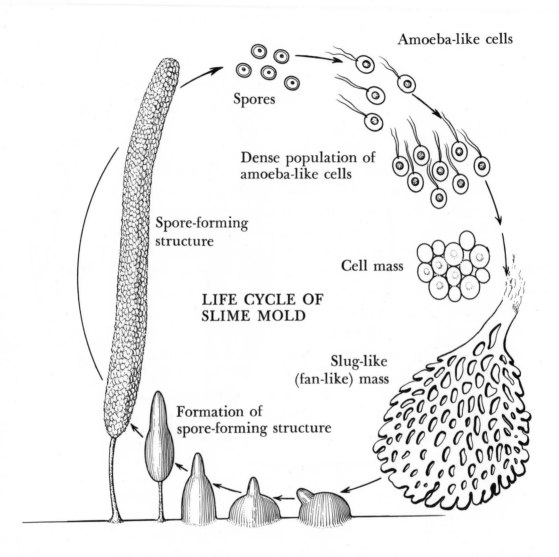

Amoeba-like cells

Spores

Dense population of
amoeba-like cells

Spore-forming
structure

Cell mass

**LIFE CYCLE OF
SLIME MOLD**

Slug-like
(fan-like) mass

Formation of
spore-forming structure

Slime Molds

One of the most unusual cell colonies is that of the slime mold. It has a life cycle of stages, going from a group of independent amoebas, feeding and growing as individuals, to a united spore-forming structure. (A *spore* is a hard case, microscopic in size, that contains a cell that can grow into a new individual.) When conditions are favorable, spore cases split open, releasing amoebas. These few cells feed and multiply rapidly. When the population becomes fairly dense and the food scarce, the amoebas all start moving toward some central point, forming a cell mass. This mass assumes a shape something like a slug, only much smaller, and starts moving in one direction. It has a definite front and hind end, and it moves as a unit. This movement is an example of remarkable coordination between individual cells.

The cells are held together with a sticky substance which is left behind as a trail when the mass moves. After a time, an incredible division of labor takes place. All of a sudden, the slime mold contracts into a ball, and some of the cells at the front end push downward into the mass of cells. The cells at the tip of this inverted cone start growing upward, forming a stalk. The cells that were at the back end are lifted up as the stalk grows. These cells become a sack of spores that will form the next generation. The cells that make up the stalk eventually die.

Cells in Higher Plants and Animals

Cell specialization reaches its highest level in the plants and animals we see when we look around us. These organisms have such efficient cells to take care of the basic requirements of other cells that they can also support those cells which are capable of doing very specialized jobs.

Let us look at what some of these specialized jobs are, and how they are performed by cells in plants and animals.

Cells for Protection

Both plants and animals contain cells that have the job of protecting other cells. Some kinds of protective cells form an outer covering that prevents underlying cells from injury, from drying out, and from the invasion of germs. Another kind of protective cell destroys germs that have somehow gotten through the covering layer.

One example of a protective cell is found on the outside of leaves. It is called *leaf epidermis*. The cells of leaf epidermis fit together like a jigsaw puzzle. The cell walls that face the outside are covered with a waxy, waterproof substance. Passage of water and gases in and out of the leaf is regulated by so-called guard cells that can open and close.

In the human body, skin cells fit together to form a continuous

Water, etc., is repelled.

Water, etc., is repelled.

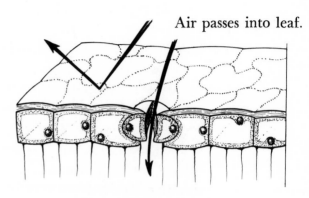

Air passes into leaf.

Cross section of leaf epidermis

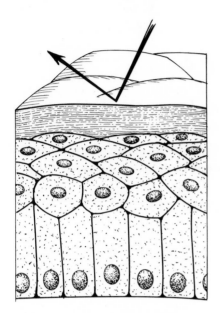

Cross section of skin epidermis

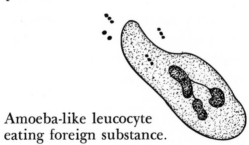

Amoeba-like leucocyte
eating foreign substance.

surface, which is covered with a protective protein made from dead skin cells. As this protein covering is constantly worn off, it is continually replaced from the underlying cells that move up and die. Can you see similarities between human skin cells and leaf epidermis? What differences can you find?

Occasionally, when germs and other foreign substances get past human skin, special cells called *leukocytes* (*lew*-ko-cites) come to the rescue. These cells move just like amoebas, and they "eat" foreign substances in the same way that amoebas engulf food. Germs inside leukocytes are made harmless to other cells. Leukocytes die when they have "eaten" many germs. Dead leukocytes collect at infections and become a part of pus.

Cells for Support

One-celled organisms do not need any structural support, because they are too small to be affected by gravity. Primitive animals such as jellyfish, and plants such as seaweed, are supported by the water in which they live. But in order for an animal to reach any substantial size on land, or to be capable of rapid movement, it must have a structure for support and for muscle attachment. Plants that have to compete with other plants for sunlight must also have some kind of support in order to grow upward.

In human beings, and in all animals with backbones, support comes from cartilage and bone. Cartilage is the flexible, smooth, translucent material found at the ends of long bones. (You can see it at the end of the breastbone of a young chicken.) The special cells that form cartilage look like tiny islands, completely surrounded by the substance of their own making.

Bone is rigid material containing minerals similar to certain minerals found in rocks. It is formed by bone cells, which are located around tiny canals that run all along bones. These canals contain nerves and blood vessels, to nourish bone cells. If you break a bone, bone cells form new bone to heal the fracture. What do you think is the job of the nerves in bone? What might happen if a broken bone did not hurt?

CARTILAGE

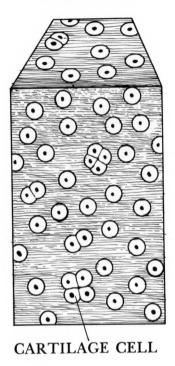

CARTILAGE CELL

BONE

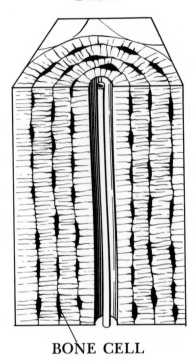

BONE CELL

Cartilage is firm
but compressible.

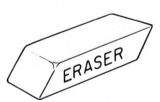

Bone is rigid like inflexible
steel tubes.

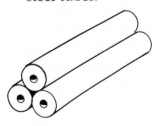

XYLEM TUBES

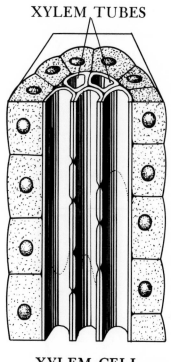

XYLEM CELL

Xylem tubes are stiff but flexible hollow tubes.

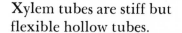

The stems of plants support and lift the leaves so that they get more sunlight. Since plant cells have rigid cell walls, plants do not need a special supporting structure as animals do. Woody plants, including trees, however, contain special cells which have particularly thick cell walls. They are called *xylem cells*. After a xylem cell has formed, the protoplasm dies, leaving behind a hollow tube. Hollow xylem cell walls are connected end to end to form long tubes. In trees, these tubes can be several yards long. In addition to support, xylem tubes conduct water and minerals up to the leaves.

Cells for Storage

Evolution of different life forms has produced animals that are prepared to survive periods when food is scarce. Grazing animals are able to guard against drought and other events that destroy vegetation. Animals that hunt other animals may have to go a long time between kills.

Protection from famine comes from food that is stored in fat cells. Each fat cell contains a fat droplet that takes up most of the space within a cell and pushes the nucleus and cytoplasm up against the cell membrane.

Plants store food in the form of starch. There are special cells in some plants, such as the potato, that contain many starch grains. In other plants, starch grains can be found in cells in roots, stems, and leaves.

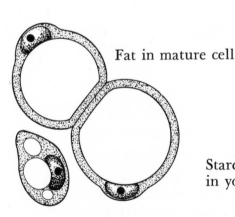

Fat in mature cell

Fat globules in
growing fat cell

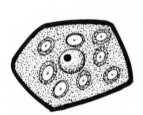

Starch in leucoplast
in young potato tuber cell

Starch in mature
plant stem cell

Cells for Movement

Motion among living things reaches its most developed form in the animal kingdom. All the kinds of motion found in one-celled plants and animals (streaming protoplasm, flagella, amoeboid motion, etc.) are found in higher animals. In addition, animals have cells that have the special job of working together to produce different kinds of motion, including the slow and continuous movement of a worm pushing through the earth and the incredibly rapid movement of a hummingbird's wings. These cells are *muscle cells*.

Muscle cells have a structure that is similar to muscles. And like muscle, they are also able to contract. In fact, the contraction of a muscle is the sum of a coordinated contraction of the cells that make up the muscle. Scientists have been very interested in the structure of the protoplasm of muscle cells. They want to find out just how muscle contraction is accomplished. They have learned that muscle cells contain tiny threads of two kinds of protein. One protein is dark and one is light. These threads are arranged in such a way that contraction results when dark protein threads move past light protein threads. In skeletal muscles, which perform the most rapid movements, there are so many of these threads that the cells have a striped appearance. Smooth muscle, found in intestines, moves very slowly, and dark and light stripes are not visible.

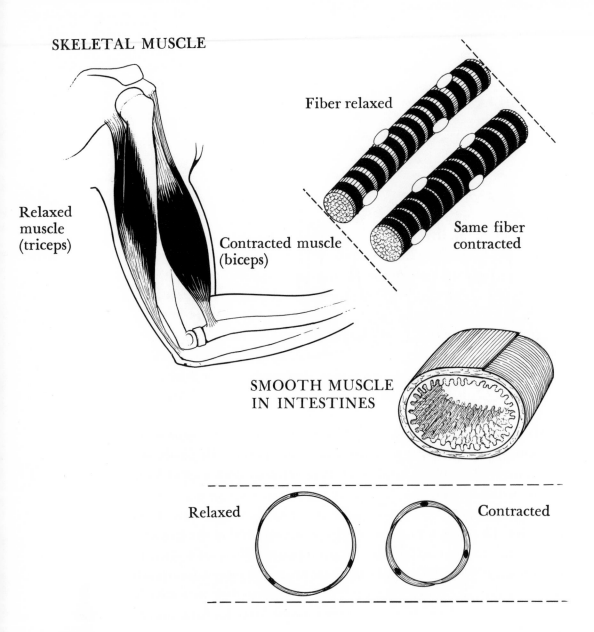

SKELETAL MUSCLE

Fiber relaxed

Relaxed
muscle
(triceps)

Contracted muscle
(biceps)

Same fiber
contracted

SMOOTH MUSCLE
IN INTESTINES

Relaxed

Contracted

CILIA

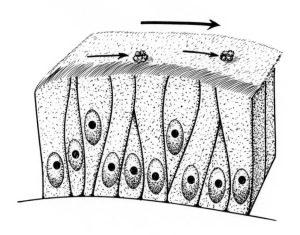

Muscle cells are not the only cells we possess that are capable of movement. In addition to leukocytes and sperm cells (male sex cells), there is a strange kind of moving cell that does a special job in our windpipes.

When we breathe in air we take in all kinds of dust and other particles. The cells lining the windpipe are covered at the free end with many hairlike extensions of protoplasm called *cilia* (*sil*-e-a). Cilia resemble flagella except that there are many more cilia in each cell, and they are much shorter. They move together in a coordinated fashion to beat dust and other foreign particles back up the windpipe.

Cells for Transportation

In large and complicated plants and animals there is the problem of transporting food and oxygen to specialized cells which are no longer able to get food and oxygen for themselves. These substances must be brought to them in a form that they can use.

The body of a higher animal has a bloodstream that continually flows through vessels that are within reach of all cells. Dissolved food molecules flow in the liquid part of the blood. Oxygen, however, is carried by special cells — the red blood cells.

Red blood cells are formed in the marrow cavities of long bones. Mature cells, released into the bloodstream, have no nuclei. They contain a protein which can combine with oxygen temporarily until cells remove it. This protein contains iron. Iron can combine with oxygen resulting in rust. (Rust forms from the chemical combination of iron and oxygen.) The iron gives the protein its red color, which becomes brighter when it combines with oxygen.

There are an enormous number of red blood cells in the human body to do the even more enormous job of bringing oxygen to every single cell. One drop of blood contains over five million red blood cells. If a person has a low red cell count or if he does not have enough iron, his cells will not get enough oxygen. This condition is known as *anemia*. Anemic people feel tired and do not have much energy.

Besides the xylem tubes that help transport water and minerals up trees, plants also have special cells, called phloem (pronounced *flome*) cells, that transport sugar and other materials to all parts of the plants. Unlike xylem cells, phloem cells remain alive. How they transport materials is still a mystery, although this activity is being studied by scientists.

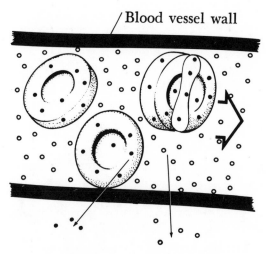

Blood vessel wall

Oxygen from
red cells

Dissolved food
from blood plasma.

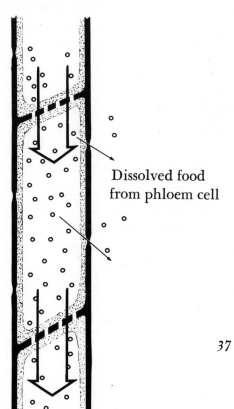

Phloem cell

Dissolved food
from phloem cell

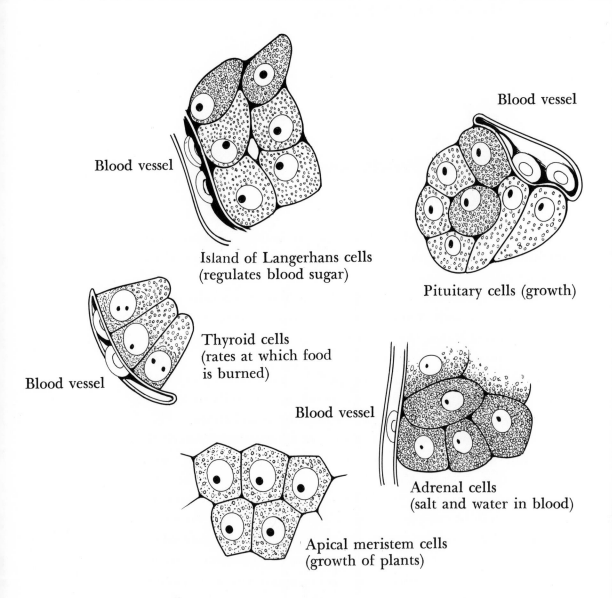

Blood vessel

Island of Langerhans cells
(regulates blood sugar)

Blood vessel

Pituitary cells (growth)

Thyroid cells
(rates at which food
is burned)

Blood vessel

Blood vessel

Adrenal cells
(salt and water in blood)

Apical meristem cells
(growth of plants)

Cells for Regulation

Both plants and animals produce chemicals that act in very small amounts to affect growth, development, and behavior. These chemicals are called *hormones*.

Plant hormones are produced in rapidly growing cells at root tips, stems, and buds. They control the rate of growth, the appearance and loss of leaves, the way leaves turn toward the sun, and the way roots turn down into the earth.

Animals produce hormones in special organs called *glands*. We have many different glands in our bodies. The cells in a gland secrete hormones which go directly into the bloodstream.

There is nothing very special about the appearance of gland cells that could give you a clue as to what their job is. Here are some drawings of different gland cells with a statement about the effect of the hormones they secrete: What can you say about the similarities between different gland cells? (Thyroid cells control the rate at which we burn food for energy; Islands of Langerhans control the amount of sugar in the blood; adrenals control emergency reaction to stress; pituitary controls growth, and interacts with other glands.)

RESPONSE TO STIMULI

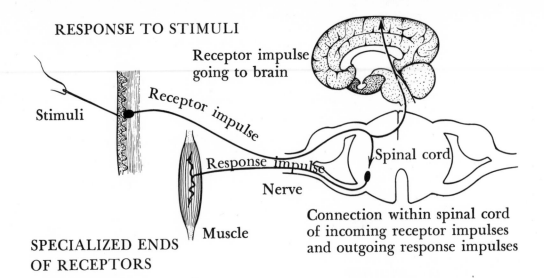

Receptor impulse going to brain

Stimuli

Receptor impulse

Response impulse

Nerve

Spinal cord

Muscle

Connection within spinal cord of incoming receptor impulses and outgoing response impulses

SPECIALIZED ENDS OF RECEPTORS

Pressure

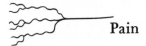

Pain

Touch

Warmth

Cold

SPECIALIZED RECEPTOR END IN EYE

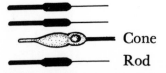

Cone

Rod

Cells for Communication

As we go up the evolutionary scale, from amoeba to man, we find organisms capable of more and more complicated interaction with environment. Amoebas respond to temperature changes and to contact with food particles. Human beings not only respond to a countless number of environmental changes (called *stimuli*), but they can act to change their environment as well. We can do this because we have special cells that receive stimuli and start muscle contraction. They are the nerve cells.

Nerve cells that receive stimuli are called *receptors*. We have many different kinds of receptors. Skin receptors respond to pain, warmth, cold, and pressure. Receptors in sense organs respond to sound, chemicals (taste and smell), light, position, and movement. Two examples of receptors are found in the eye. Rod cells are sensitive to dim light and are responsible for black and white vision and "night" vision. Cone cells are sensitive to all the shades in the rainbow, and they are the reason that we see colors.

A receptor responds to a stimulus by sending an "impulse" along many connecting nerves to the brain or spinal cord. Nerves are made of bundles of nerve cells. The brain and spinal cord act as a giant switchboard or computer. Connections between the incoming impulse and nerves leaving the brain or spinal cord are made here. Outgoing impulses cause muscles or glands to make a response.

The nerve impulse is a subject of much scientific investigation. It has electrical properties and can be picked up by electrical instruments. (You may have heard of brain waves that are recorded during hospital examinations.) It is not, however, like the electricity that travels through wires. Scientists hope to learn more about these mysterious impulses by studying them in the especially large nerve cells found in squids. They study squid nerves because there is no question that the principles that account for the way in which impulses travel along a squid nerve are the same for nerve cells in all animals.

Reproduction in Cells

So far we have discussed only one aspect of what is called modern cell theory; namely, that all living things are made of cells. But there is another important general statement that is also a part of modern cell theory. In 1858, a German biologist named Rudolf Virchow (1821–1902) stated, "All cells come from cells."

This simple statement has two implications: (1) All present living things come from preexisting living things: organisms do not arise from nonliving material at this time in evolution; and (2) Since life passes from one generation to the next in a continuous way, all present forms of life must have descended from common ancestors.

How, then, do cells come from cells? Most cells reproduce by dividing into two "daughter" cells that are identical to the parent cell and to each other. Cell division is an extremely orderly process in which the nucleus first duplicates its contents with mathematical precision. Just before division these contents become visible under the microscope as they become organized into dark staining bodies called *chromosomes*. The chromosomes, which are in pairs, line up in the middle of the cell. They then split up as each member of the pair migrates to opposite ends of the cell. Each group forms a new nucleus, and a new cell membrane is formed to divide the rest of the cytoplasm. The division of the cytoplasm is not always as exact as the division of the nucleus.

This process of cell division, called *mitosis* (my-*toe*-sis), is of extreme importance to the continuity of life. Through mitosis each cell transmits its particular characteristics from one generation to the next. The "plans" that determine whether a cell is a skin cell,

MITOSIS

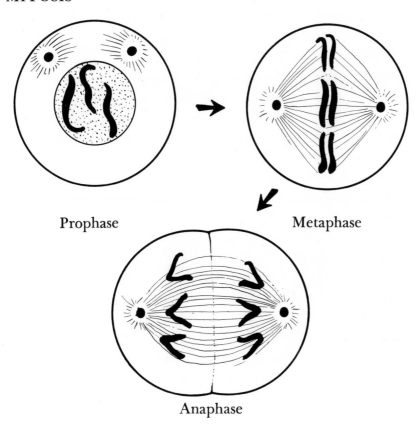

Prophase

Metaphase

Anaphase

muscle cell, or phloem cell are found in the chromosomes of the parent cell.

Cells that undergo mitosis most often are not very specialized. Which cells would you guess are least likely to reproduce themselves? If you've guessed nerve cells you would be right. Young plants and animals that are still actively growing contain many cells in the process of mitosis.

MITOSIS

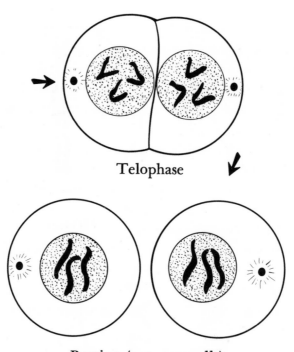

Telophase

Resting (two new cells)

Cells for Reproduction

Reproduction in one-celled organisms is very straightforward. Cell division results in two cells that can eventually divide and produce four cells and so on. In simple cell colonies, like the volvox, only the cells at the back end of a volvox are responsible for forming new volvox cells and colonies. The forward cells do not reproduce.

In higher plants and animals, reproductive cells have become even more specialized. There are, first of all, two kinds of reproductive (or sex) cells — male and female. In plants, male reproductive cells are found in pollen grains. Female cells are called *ova* and are found in special structures near the base of flowers. In animals the male sex cells are called *sperm* cells and the female are called *egg* cells. Any *one* of these sex cells cannot become a new individual all by itself. It must unite with a cell of the opposite sex in order for a new generation to develop. The process in which these reproductive cells unite is called *fertilization*.

Fertilization gives to a species a survival value that is not found among organisms that have one parent. The daughter cells formed from a single parent cell are identical to the parent cell. That is, they inherit both the good and bad traits possessed by the parent.

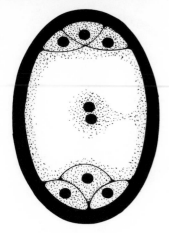

PLANT

Female ovum

Sex cells have exactly half the number of chromosomes as do the cells of the parent. And the half that each cell contains is determined by chance. But the full number of chromosomes is restored by fertilization. This means that the offspring with two parents will have some of the traits of both and will not be identical to either one. There is always a chance that a new individual will inherit the worst traits of both parents. If it does, it will probably not survive to reproduce itself and pass these traits on. In this way, slowly, the worst traits of a species disappear, and the species, as a whole, improves.

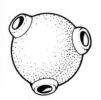

Various pollen grains (male)

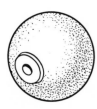
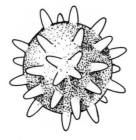

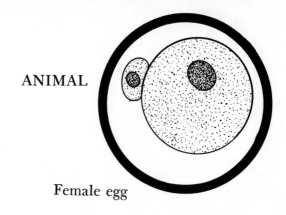

ANIMAL

Female egg

There are many ingenious methods in nature for making certain that fertilization will occur. Some pollen grains grow tubules to the ova to make sure the male cells will reach their target. Sperm cells of many animals are equipped with flagella that enable them to swim to an egg. Fertilization in animals may take place outside the parents' bodies, as with fish and insects. Or it may be internal, as is the case with birds and man.

After fertilization, a new organism begins to develop. At first, the fertilized egg goes through division that produces more and more unspecialized cells. After a while, the unspecialized cells begin to

Protective covering

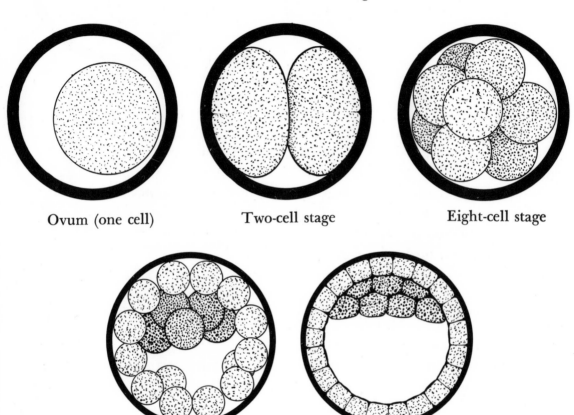

Ovum (one cell)   Two-cell stage   Eight-cell stage

Beginning of Differentiation

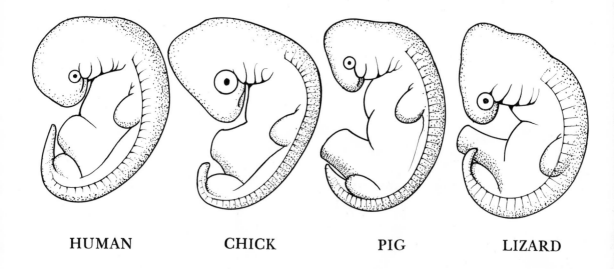

HUMAN   CHICK   PIG   LIZARD

change into the different kinds of cells you have just read about. This process of change is called *differentiation.*

The causes of differentiation have been a subject of intense scientific study. These causes are a mystery because we cannot see with our best microscopes what happens to change an unspecialized cell into a nerve cell, a muscle cell, or a skin cell. The causes must be chemical, and we are only now beginning to find evidence of such chemicals.

One striking fact about the development of new life from fertilized eggs is that all developing life shows a similar pattern as it changes from a single fertilized egg to a fully formed embryo (*em-bree-o*) — the stage of development prior to birth. At early stages of development you cannot tell the difference between a future frog, chicken, pig, or human. And at different times a new human being resembles a simple cell colony, a jellyfish, a worm, a fish, and a bird.

49

The Diseased Cell

Cancer is probably the most dreaded of all diseases today. If untreated, or detected too late, it can cause a painful death, with the victim growing weaker day by day. It can progress quickly or it can be very slow.

Cancer is a disease of cell division, and it can occur in almost every kind of cell. There is cancer of the skin, muscle, brain, and so on. It is thought to be caused by a variety of factors, including cigarette smoke, chemicals, viruses, and penetrating radiation. Somehow, all these different "causes" do one thing; they throw a "monkey wrench" into the orderly machinery of cell division. Scientists have yet to discover just how and why cancer comes to exist. When they do, they will have the key to the true cause of cancer.

Although cancer may be caused by many factors, and no matter what kinds of cells become cancer cells, all cancer cells have three things in common: (1) Cell division does not proceed in the orderly fashion of mitosis found in normal cells; in cancer cells, cell division is wild and disorganized. (2) No matter how specialized the original cells were, cancer cells have the appearance of unspecialized cells. They seem to have gone through a process that is the reverse of cells in a developing embryo. In an embryo, unspecialized cells become specialized. With cancer, specialized cells become unspecialized. (3) Cancer cells can migrate through the bloodstream and grow in any part of the body. This is why it is important to detect cancer early, before it begins to spread.

Cancer is treated by surgery, drugs, and penetrating radiation. But there are problems. Sometimes surgeons cannot cut out all of the cancer because to do so would mean removal of, or damage to, a vital organ, which would cause the death of the patient anyway.

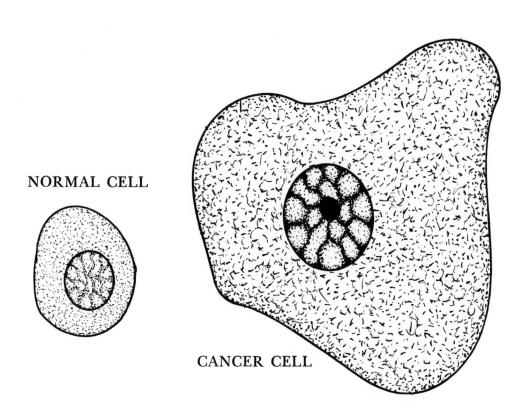

NORMAL CELL

CANCER CELL

Therefore, some cancers must be treated with drugs or radiation. For some reason, rapidly growing cells of cancer are more susceptible than normal cells to the killing effects of radiation and drugs. But drugs and radiation can damage normal cells in treatments designed to kill only cancer cells. Ironically, some drugs and radiation, used to cure cancer, can also cause cancer.

The real cure for cancer will come when we know the answers to basic questions about all cells. What controls cell division? What controls cell growth and differentiation? How do causes of cancer interfere with normal cell division? What is the basic difference in the chemistry of normal cells and cancer cells? Why do normal body defenses fail to act against cancer as they react to other diseases?

The Electron Microscope:
A Tool to Probe a New World

In their search to find the answer to the causes of cancer and other questions, scientists have developed a new tool. It is a microscope that uses tiny, electrically charged particles instead of light to "see" specimens. It is called the "electron" microscope. This remarkable and complicated instrument (which takes up half a room) has opened up as new a world to scientists as the light microscope did over 250 years ago. Although the best light microscope can magnify a cell to 2,000 times actual size, the electron microscope clearly reveals detail at 200,000 times actual size!

75 times actual size   25,000 times actual size   100,000 times actual size

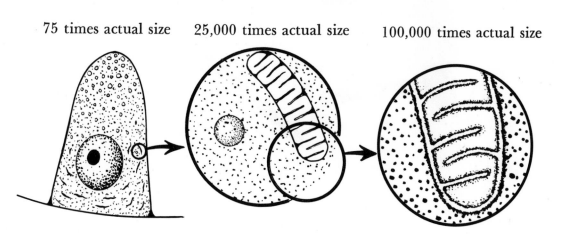

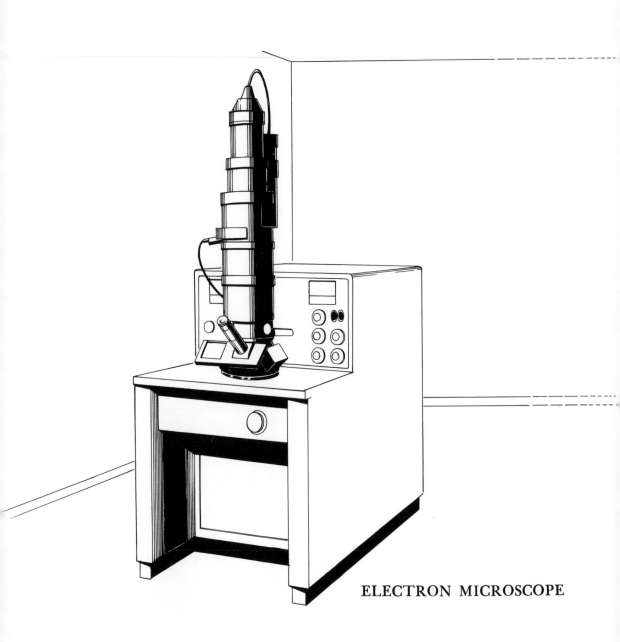

ELECTRON MICROSCOPE

At such great magnification a cell looks very different from cells under a light microscope. Protoplasm is not a formless blob but a highly organized material. Perhaps what is even more striking is that the differences between all kinds of cells are less noticeable and important than their similarities. In other words, it would be difficult to tell the kind of cell that a particular sample of protoplasm came from. All protoplasm is remarkably alike.

The secrets of life have been studied in different ways. Some scientists study intact organisms, others study cells under a microscope, still others study interactions between the molecules of life. There is hope that the electron microscope will help to answer many questions by bridging the gap between information obtained through light microscopes and information obtained from chemistry laboratories. If this hope is realized and scientists find answers to today's questions, there can be no doubt that these answers will raise even more questions to be answered in the future.

GLOSSARY

AMOEBA — a one-celled animal without any definite shape.

ANEMIA — condition in which blood does not carry enough oxygen.

CARTILAGE — flexible, smooth material found at ends of certain bones.

CHLOROPHYLL — green pigment in plants.

CHROMOSOMES — dark staining bodies which appear in cell nuclei just before cell division.

CILIA — short hairlike extensions of protoplasm that move in a co-ordinated way.

CONTRACTILE VACUOLE — empty-looking body in an amoeba used to expel wastes.

CYTOPLASM — cellular protoplasm except the nucleus.

DIATOMS — one-celled green plants with glasslike skeletons.

DIFFERENTIATION — process of change where unspecialized cells become specialized.

EGG CELLS — female sex cells.

ELECTRON MICROSCOPE — instrument that uses tiny, electrically charged particles to magnify specimens up to 200,000 times actual size.

EMBRYO — early stage of development of young from a fertilized egg.

FERTILIZATION — process of uniting egg and sperm cells.

FLAGELLA — long, whiplike extensions of protoplasm used to propel cells through fluids.

FOOD VACUOLE — empty-looking body in one-celled animals where food is digested.

GLANDS — special organs in the body whose cells secrete chemicals.

HORMONES — chemicals in plants and animals that affect growth, development, and behavior.

LEAF EPIDERMIS — protective cells on outsides of leaves.

LEUKOCYTES — cells which protect against germs and foreign substances that get past skin.

MICROSCOPE — instrument that enlarges the image of a natural object.

MITOSIS — orderly process of cell division where the nucleus divides before the cytoplasm.

MUSCLE CELLS — cells which work together to produce motion.

NUCLEUS — central body of protoplasm of most cells.

ORGANISM — any living thing.

OXIDATION — combining food (or any fuel) with oxygen to produce energy.

PHLOEM CELLS — cells which transport sugar and other materials to all parts of plants.

PROTOPLASM — jellylike living substance found in all cells in both plants and animals.

PROTOZOA — group of one-celled animals.

PSEUDOPODS — "false feet" of an amoeba used for locomotion and obtaining food.

RECEPTORS — nerve cells that receive stimuli.

SPERM CELLS — male sex cells.

SPORE — hard case containing cell that can grow into new individual.

STIMULI — environmental changes.

VOLVOX — colony of cells that moves in a rolling fashion.

XYLEM CELLS — thick-walled hollow cells found in plants, used for support and water transport.

INDEX